LIVING COLORS:

Microbes of Yellowstone National Park

In Yellowstone's thermal areas, stay on the boardwalks and designated trails to protect yourself and preserve natural areas. Do not touch thermal features. You could be severely burned.

©2013 Montana State University
All rights reserved.

Montana State University
304 Montana Hall
P.O. Box 172460
Bozeman, MT 59717-2460
Phone: 406-994-7868

ISBN: 0-934948-22-4

CONTRIBUTORS:
Eric Boyd, Monica Brelsford, Jamie Cornish, John Dore, Jerry Johnson, Susan Kelly, John Peters, Brent Peyton, Heather Rauser, Christine Smith, Suzi Taylor, Jennifer Wirth, Mark Young

DESIGN AND LAYOUT:
Monika Chodkiewicz

IMAGES:
Darren Edwards: Dominant photo on cover, inside front cover, 1, 6, 14, 16, 20, 22, 24, 30, 40-41, 42-43, 44-45, 46-47; Inserts on pages 10, 12, 20, 36, 38
David Patterson: Inserts on pages 4, 6, 14, 16, 18, 24, 30, 32
National Park Service: Dominant photo on pages v, 8, 12, 18, 28, 36
Marla Goodman: Inserts on pages 2, 8, 26, 28, 34
Thermal Biology Institute: Dominant photo on page 2, 4, 10, 26, 32, 34, 38, inside back cover, back cover; Insert on page 22
Monika Chodkiewicz: Illustrations on pages 48, 50, 51

PUBLISHED BY YELLOWSTONE ASSOCIATION

Visit tbi.montana.edu/livingcolors to view animations of microbes featured in this book.

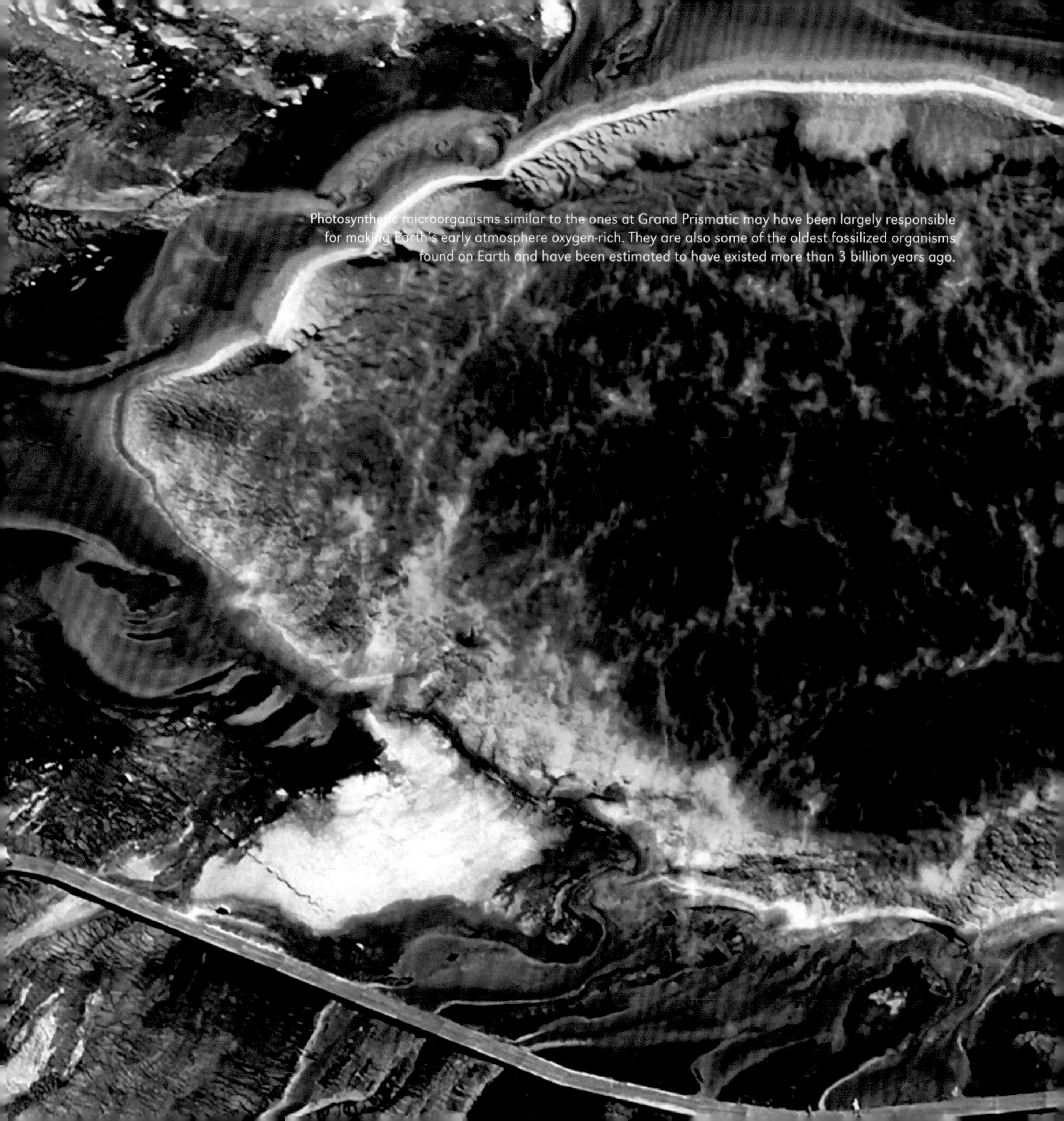

Photosynthetic microorganisms similar to the ones at Grand Prismatic may have been largely responsible for making Earth's early atmosphere oxygen-rich. They are also some of the oldest fossilized organisms found on Earth and have been estimated to have existed more than 3 billion years ago.

Introduction

Millions of people visit Yellowstone every year to view its majestic sights and wild animals — but we often overlook the billions of microscopic creatures under our feet. In fact, there may be more microbes in the soil directly under our feet than there are people living on Earth, but why should we care? Of what importance are these tiny organisms?

Microbes are individual living organisms smaller than our eyes can see. They are the oldest form of life on Earth, and they live just about everywhere, including in, and on, your body. Some kinds of microbes, called extremophiles, can thrive in places that are freezing cold, super hot, deep underground, or at the bottom of the ocean. These environments that are considered "extreme" on Earth are similar to what is "normal" on other planets or moons. Therefore, scientists think if we find life elsewhere in the universe it may resemble the life on Earth living in extreme environments.

The extremophiles of Yellowstone are incredibly important for many other reasons. The rainbows of microbial life swirling in and around the hot springs of Yellowstone have already yielded the secret for how to better amplify DNA in the laboratory, which impacts several areas of biotechnology such as helping solve crimes, identifying crime victims, and playing a key role in medical diagnosis and treatment. Scientists are now actively pursuing how these extremophiles might also be a source for alternative fuels such as hydrogen, a means of improving medical cures for diseases such as cancer, or mechanisms for cleaning hazardous waste sites. Yellowstone's microbes may also offer designs or processes that engineers can mimic to solve human problems.

Microbes play a critical role in all nutrient cycles. Two important examples are processes that are essential to making Earth capable of sustaining life – the carbon cycle and the nitrogen cycle. The nitrogen cycle is important for food production and plant growth. The carbon cycle impacts the abundance of oxygen and greenhouse gases and can alter climates.

Microbes can even help shape the landscape. For example, at Roaring Mountain in Yellowstone, you can see *Sulfolobus acidocaldarius* eating away at the mountain. These microbes consume the rotten egg-smelling hydrogen sulfide gas escaping from deep below the ground and convert it into sulfuric acid, which dissolves rocks and turns them into mud.

Even the viruses that infect Yellowstone microbes are proving to be incredibly valuable. The shells of the viruses, or protein cages, can be used as containers for metals or medicines to enhance targeted drug delivery, improve MRI imaging, or even increase computer memory.

When you take all of these scientific discoveries into account, the microbes of Yellowstone offer us a means for learning more about ourselves – from our origins to prolonging our lives through biotechnological and biomedical advances.

Yellowstone is an awe-inspiring place to visit for anyone interested in microbes, because it is one of the few places in the world where an amazing diversity of microbes can be seen without a microscope. Large microbial mats and filaments with distinct white, yellow, green, pink, orange, brown, and black colors are visible to the naked eye. Yellowstone is also a particularly rich environment in which to explore extreme life, because it is home to half (more than 10,000) of the world's hydrothermal features. Yellowstone's thermal features have a wide range of pHs, temperatures, and chemical compositions that yield a diversity of microbes that we are just beginning to discover.

Although the larger flora and fauna of Yellowstone may be more obviously charismatic and visible, the microbes of Yellowstone should not be overlooked. This book was created as a sort of "wildlife guide" to introduce visitors to a few of the microbes of the Park. The microbes represented in this book do not encompass all of the species that visitors will encounter, but they are some of the more prevalent and well-understood.

Chlorobium

MAMMOTH HOT SPRINGS

Chlorobium

domain: **BACTERIA**
phylum: **CHLOROBI**

pH

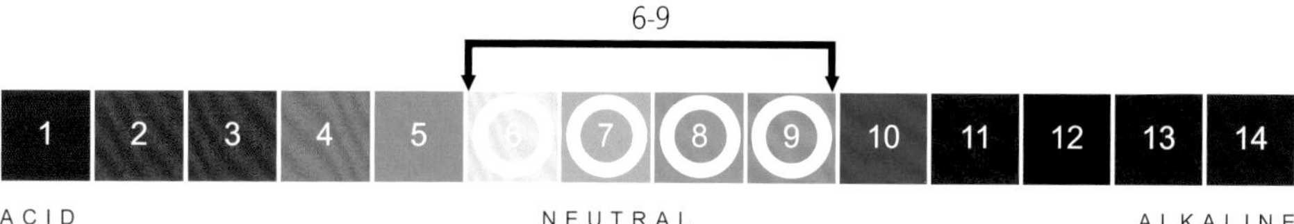

Temperature

32-52°C (90-126°F)

Description

Chlorobium is a rod-shaped green sulfur bacterium that often grows in long chains to form dense mats. It is typically dark green, conducts photosynthesis without making oxygen and produces sulfur, or more commonly sulfate, as a waste product. Some scientists think species of *Chlorobium* may have played a part in a few of the mass extinction events on Earth. This is because during periods when oceans were lower in oxygen, *Chlorobium* would have been more successful than other photosynthesizers and would have produced large amounts of methane and hydrogen sulfide, therefore increasing global temperatures and acid rain.

Habitat

Chlorobium is also found at Calcite Springs (the downstream end of the Grand Canyon of the Yellowstone) and at Thermopolis Springs in Wyoming. Species of *Chlorobium* have been found in hot springs in Eastern Oregon, Nevada, New Mexico, California, and New Zealand.

Chloroflexus

MAMMOTH HOT SPRINGS

Chloroflexus

domain: **BACTERIA**
phylum: **CHLOROFLEXI**

pH

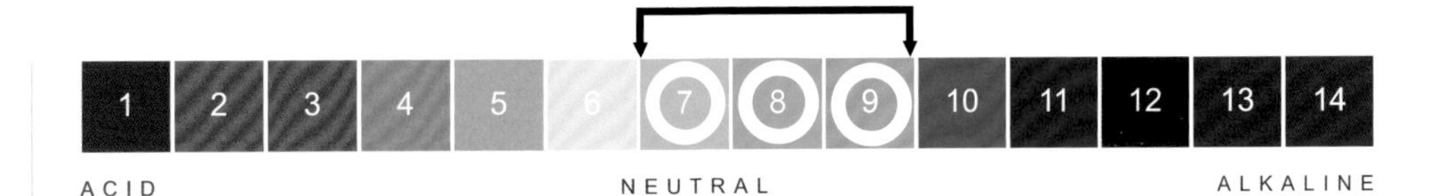

Temperature **35-85°C (95-185°F)**

Description

Chloroflexus is a filamentous anoxygenic phototroph, formerly referred to as a green nonsulfur bacterium, that is rod shaped and forms filaments. It uses light for energy but uses organic carbon derived from other organisms to make new cell material and does not produce oxygen as a byproduct of photosynthesis. In spring waters that contain sulfur, it may be the first green-colored mat that you will see. Scientists are studying ***Chloroflexus*** because they think it may shed light on the evolution of photosynthesis.

Habitat

Chloroflexus is found in various thermal springs around Yellowstone – such as in runoff from Octopus Spring, along Firehole Lake Drive, in Mammoth Hot Springs, and in many alkaline springs on the Lower, Midway, and Upper Geyser Basins.

Oscillatoria

MAMMOTH HOT SPRINGS

Oscillatoria

domain: **BACTERIA**
phylum: **CYANOBACTERIA**

pH

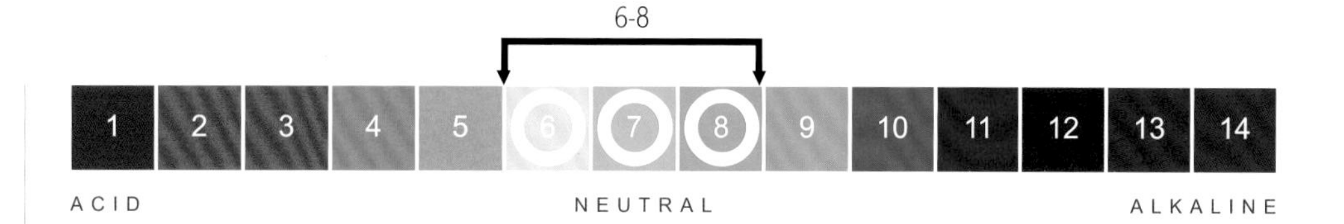

Temperature

36-45°C (96-113°F)

Description

Named for its slow back and forth oscillating movement that allows it to move closer to a light source, *Oscillatoria* is a rod-shaped bacterium that often forms mats. Scientists think it moves by releasing a glue-like substance that propels it in the opposite direction as the secretion it makes. *Oscillatoria* tends to be orange and produces energy through photosynthesis. Some species may also fix nitrogen, a process that converts nitrogen gas (N_2) into a chemically reactive form of nitrogen that plants can use in photosynthesis. Enzymes responsible for this process are vulnerable to destruction by oxygen, and nitrogen fixation is often performed in environments that lack oxygen. To create an anaerobic (oxygen-free) environment, this organism forms a tight ball of bacterial filaments that keep dissolved oxygen away from the nitrogen-fixing enzymes, which are extremely sensitive to oxygen.

Habitat

Oscillatoria species can be found in hot springs around Yellowstone below 45°C (113°F) such as Mammoth Hot Springs and the Chocolate Pots. *Oscillatoria* species have been isolated all over the world on every continent except Antarctica.

Sulfurihydrogenibium

MAMMOTH HOT SPRINGS

Sulfurihydrogenibium

domain: **BACTERIA**
phylum: **AQUIFICAE**

pH

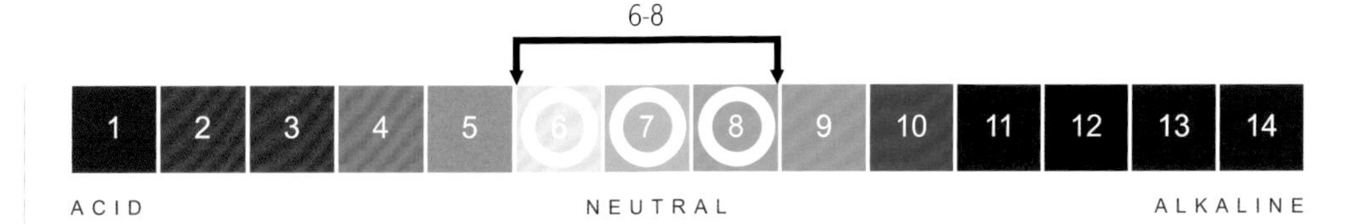

Temperature

60-75°C (140-167°F)

Description

Sulfurihydrogenibium species are straight to slightly curved rods of bacteria. They often form cream filaments or streamers and precipitate iron and sulfur. *Sulfurihydrogenibium* often live on elemental sulfur, thiosulfate, ferrous iron or hydrogen as an electron donor and oxygen as an electron acceptor.

Habitat

Sulfurihydrogenibium can be found around Yellowstone in places such as Mammoth Hot Springs, Calcite Springs, and Obsidian Pool. Species have been found living in hot springs located in Iceland, Russia, and the Archipelago of the Azores. They have also been found in deep-sea hydrothermal vents and even in a hot subsurface aquifer in a Japanese gold mine.

Thermochromatium

MAMMOTH HOT SPRINGS

Thermochromatium

domain: **BACTERIA**
phylum: **PROTEOBACTERIA**

pH

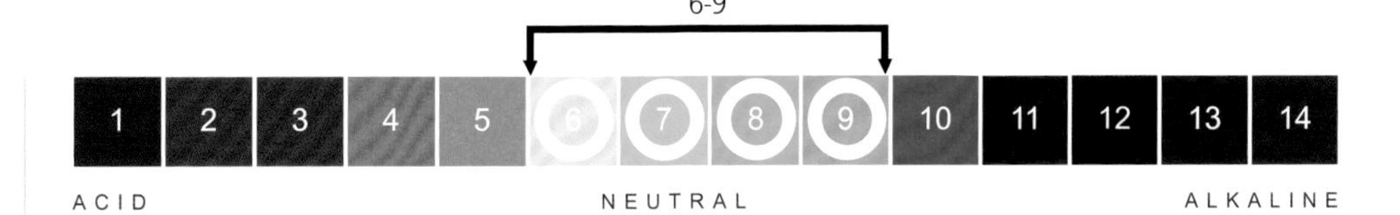

Temperature

34-57°C (93-135°F)

Description

Thermochromatium is usually deep purple red in color but can be reddish brown as well. This spherical bacterium performs photosynthesis, uses hydrogen sulfide (H_2S) in the process, and produces sulfate as a byproduct. Where large amounts of hydrogen sulfide accumulate it can form massive blooms.

Habitat

Thermochromatium has been found in a few small springs in the Mammoth area of Yellowstone, and is quite abundant in some of the hydrogen sulfide-containing springs at Thermopolis, Wyoming. Similar thermophilic (heat loving) purple bacteria have been found in hot springs in New Mexico and can be present in sewage.

Caldisphaera

NORRIS GEYSER BASIN

Caldisphaera

domain: **ARCHAEA**
phylum: **CRENARCHAEOTA**

pH

Temperature **65-75°C (149-167°F)**

Description *Caldisphaera* grows in high-temperature and acidic environments; thus it is known as a thermoacidophile. It uses organic forms of carbon and elemental sulfur as sources of energy and will convert yellow precipitated sulfur into hydrogen sulfide – a gas that smells like rotten eggs and is very poisonous and corrosive. *Caldisphaera* cells are spherical and are found singly or in pairs.

Habitat *Caldisphaera* is widely distributed in acidic hydrothermal features in Yellowstone, such as Monarch Geyser in Norris, where it is found in the yellow sulfur that is often present. *Caldisphaera* species have been found in hot springs in the Philippines, Russia, and California.

Cyanidioschyzon

NORRIS GEYSER BASIN

Cyanidioschyzon

domain: **EUKARYOTA**
phylum: **RHODOPHYTA**

pH

0-4

1	2	3	4	5	6	7	8	9	10	11	12	13	14

ACID NEUTRAL ALKALINE

Temperature **40-55°C (104-131°F)**

Description

Cyanidioschyzon is a spherical green-colored red alga found in acid springs. It uses sunlight for energy, and performs oxygen photosynthesis by processes identical to those in cyanobacteria and plants. *Cyanidioschyzon* is one of the most heat- and acid-tolerant algae known.

Habitat

Cyanidioschyzon is found in acidic hot springs around Yellowstone such as Lemonade Creek, Nymph Creek and in Norris Geyser Basin.

Euglena

NORRIS GEYSER BASIN

Euglena

domain: **EUKARYOTA**
phylum: **EUGLENOZOA**

pH

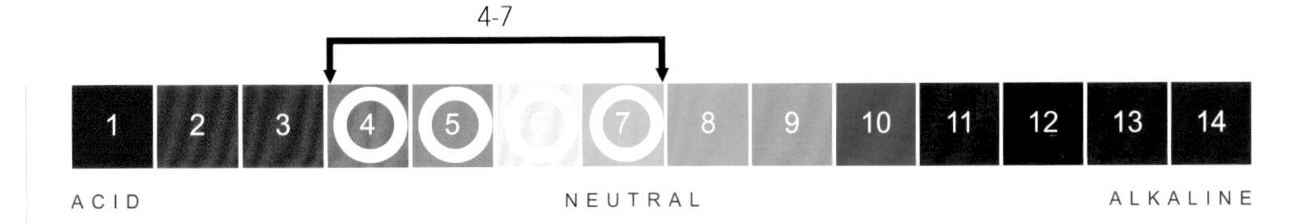

Temperature **ambient-40°C (ambient-104°F)**

Description

Euglena is a free-moving single cell organism that displays characteristics of both animals and plants. It can perform photosynthesis for energy like plants or eat similarly to animals. *Euglena* has a whip-like structure that helps propel the cell through the water. It is green with a small red eyespot that filters light before it reaches a photoreceptor in its cell. The photoreceptor helps it orient itself towards or away from light. *Euglena* can live in acidic environments but is also commonly found in freshwater streams and lakes. They often multiply into large enough groups that they form a green layer on water.

Habitat

Euglena can be found around Yellowstone in areas such as Lemonade Creek, Nymph Creek and Beaver Lake. Species of *Euglena* can be found in freshwater ponds and streams around the world.

Hydrogenobaculum

NORRIS GEYSER BASIN

Hydrogenobaculum

domain: **BACTERIA**
phylum: **AQUIFICAE**

pH

Temperature **55-72°C (131-162°F)**

Description

Hydrogenobaculum is a rod-shaped bacterium that can form yellow and white streamers. It uses hydrogen, hydrogen sulfide, and carbon dioxide for energy. It has also been shown to use arsenic as a source of energy, but only in the absence of hydrogen sulfide gas. Species of *Hydrogenobaculum* are thermoacidophiles, meaning they live in hot acid.

Habitat

Hydrogenobaculum is found in Yellowstone in acidic streams and pools in places such as Amphitheater Springs and Norris Geyser Basin. It was first isolated in hot springs located in Japan.

Metallosphaera

NORRIS GEYSER BASIN

Metallosphaera

domain: **ARCHAEA**
phylum: **CRENARCHAEOTA**

pH

2-4

1	2	3	4	5	6	7	8	9	10	11	12	13	14

ACID NEUTRAL ALKALINE

Temperature **50-80°C (122-176°F)**

Description

Metallosphaera is a spherical-shaped member of the domain Archaea that appears orange when in large groups. Like many Archaea, this organism is a hyperthermophile, which means it can grow at very high temperatures, much higher than most organisms can withstand. It uses iron / sulfur compounds such as pyrite (fools' gold) in the presence of oxygen to produce energy for metabolism and may use carbon dioxide or organic carbon to build cell structures. *Metallosphaera* is named for its ability to dissolve metals.

Habitat

Metallosphaera is found in acidic springs around Yellowstone such as Whirligig Geyser in Norris Geyser Basin. Similar species have been found in thermal areas in Italy and in acidic mine drainages, such as a slagheap of a uranium mine in Germany.

Sulfolobus
NORRIS GEYSER BASIN

Sulfolobus

domain: **ARCHAEA**
phylum: **CRENARCHAEOTA**

pH

2-4

1	2	3	4	5	6	7	8	9	10	11	12	13	14

ACID NEUTRAL ALKALINE

Temperature **65-87°C (149-188°F)**

Description

Sulfolobus species have spherical cells with lobes, and they metabolize sulfur or sulfur compounds, thus earning the name *Sulfolobus*. This organism was first isolated from Congress Pool in Norris Geyser Basin by Thomas Brock in 1972. It was one of the first hyperthermophiles –organisms that optimally grow above 80°C or 176°F – and one of the first Archaea – a domain of single-celled microorganisms with no cell nucleus – to be discovered.

Habitat

Sulfolobus has been found in acidic streams and pools around Yellowstone in areas such as Mud Volcano and Norris. Species of the genus *Sulfolobus* have also been found in sulfurous volcanic areas or hot springs on Mount St. Helens, and in Italy, Russia, Chile, Japan, and Papua New Guinea.

Zygogonium

NORRIS GEYSER BASIN

Zygogonium

domain: **EUKARYOTA**
phylum: **CHLOROPHYTA**

pH

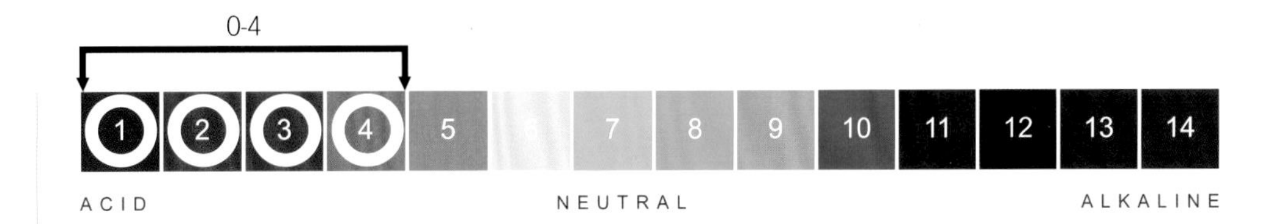

Temperature **32-55°C (90-131°F)**

Description

Zygogonium is a green rod-shaped alga that obtains its energy by performing photosynthesis in the same manner as plants. When it is exposed to intense sunlight, a dark purplish pigment is formed within its cell. Therefore, these purple cells can look almost black. These algae form thick mats near acidic hot springs and their runoff channels. The mats have a high water-holding capacity and create a moist habitat for *Euglena*, a mobile single-celled organism, and for acid-tolerant soldier fly larvae. *Zygogonium* is often found adjacent to the bright green, but more thermotolerant *Cyanidioschyzon* mats. Being less thermotolerant, *Zygogonium* species inhabit lower temperature environments than *Cyanidioschyzon*. This distinction often creates a stark boundary between mats, such as is observed in the adjacent photo.

Habitat

Zygogonium is found around Yellowstone in acidic springs such as those found at Nymph Lake, Norris Geyser Basin and Lemonade Creek. Species of *Zygogonium* also have been discovered all over the world in locations such as Canada, Australia, South Africa and India.

Calothrix

UPPER, MIDWAY,
LOWER GEYSER BASINS

Calothrix

domain: **BACTERIA**
phylum: **CYANOBACTERIA**

pH

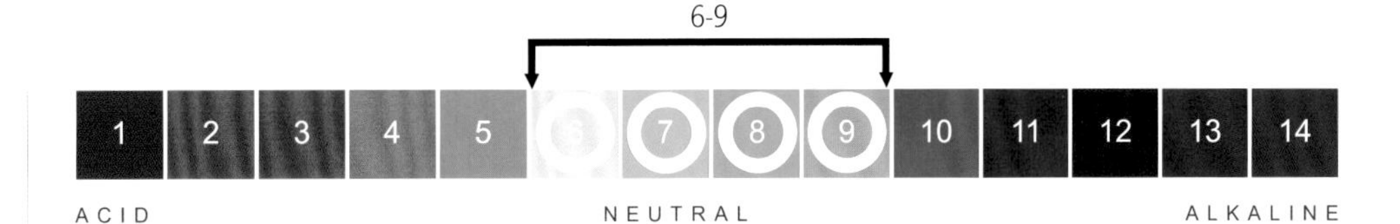

Temperature **30-45°C (86-113°F)**

Description

Calothrix is a brownish rod-shaped cyanobacterium that forms dark brown mats along the moist edges and in the outflow of many thermal features. It contains a pigment that acts like a sunscreen and protects *Calothrix* from high levels of UV radiation. It has two very different metabolisms – using photosynthesis during the day for energy and fermentation at night. Cyanobacterial ancestors of *Calothrix* are thought to be the first oxygen-producing organisms on Earth and potentially responsible for changing Earth's oxygen-poor atmosphere to a more oxygen-rich atmosphere. *Calothrix* is also a nitrogen fixer, which means that it can convert nitrogen (N_2) in the air to a useable form of nitrogen for itself and other organisms.

Habitat

Calothrix is found around Yellowstone in association with almost all hot springs within the pH range of 6-9 – including Upper, Midway, and Lower Geyser Basins, Grand Prismatic Spring and Mammoth. Species of *Calothrix* have been discovered in locations around the world including Iceland, the Gulf of Mexico, India, and Slovakia.

Desulfurococcus

UPPER, MIDWAY, LOWER GEYSER BASINS

Desulfurococcus

domain: **ARCHAEA**
phylum: **CRENARCHAEOTA**

pH

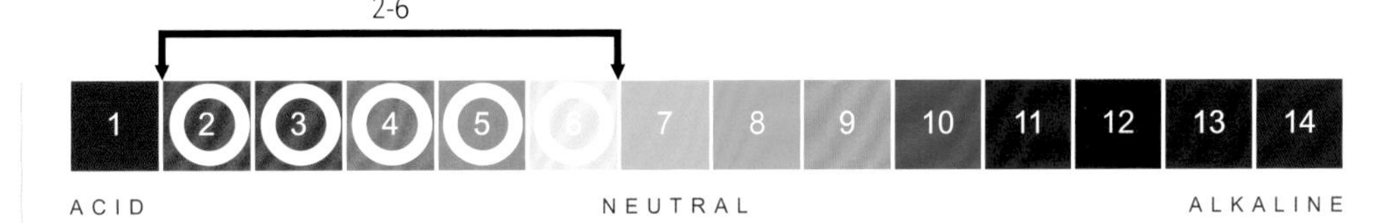

Temperature

70-106°C (158-223°F)

Description

Desulfurococcus is spherical with a surface protein that forms a lattice of mesh cross-shaped units. It does not need oxygen but obtains energy by ingesting organic carbon compounds like sugars and lipids. *Desulfurococcus* is very important because it contains an enzyme that is useful in genome engineering and gene therapy. The enzyme can change a genomic sequence in several ways and can cleave unwanted viral DNA.

Habitat

Found in Obsidian Pool, Sylvan Springs and Lower Geyser Basin in Yellowstone, *Desulfurococcus* species have also been found in Iceland and Russia.

Phormidium

UPPER, MIDWAY,
LOWER GEYSER BASINS

Phormidium

domain: **BACTERIA**
phylum: **CYANOBACTERIA**

pH

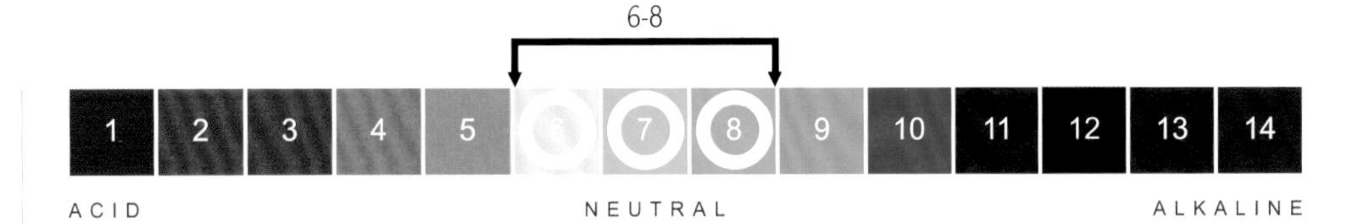

Temperature

35-57°C (95-135°F)

Description

Phormidium is a rod-shaped cyanobacterium that forms bacterial mats and performs photosynthesis for energy. It is typically orange in color and can form filaments or stromatolites – layered sedimentary structures. Several species of *Phormidium* are being studied as potential sources of toxins harmful to humans that may also be of use in shrinking or destroying tumors. *Phormidium* can be used as a biomarker (signature) to identify extinct hot springs and is therefore of interest to NASA with regard to searching for life on other planets and moons.

Habitat

Phormidium is found around Yellowstone in features such as Octopus Spring, Grand Prismatic Spring, and Queen's Laundry Spring, and in some thermal features in Mammoth Hot Springs. Species of *Phormidium* are also found in various sites around the world, such as hot springs in Chile and Turkey. There is a species of *Phormidium* that lives in the Antarctic, can withstand low temperatures, high UV radiation, and long periods without light or nutrients.

Synechococcus

UPPER, MIDWAY,
LOWER GEYSER BASINS

Synechococcus

domain: **BACTERIA**
phylum: **CYANOBACTERIA**

pH

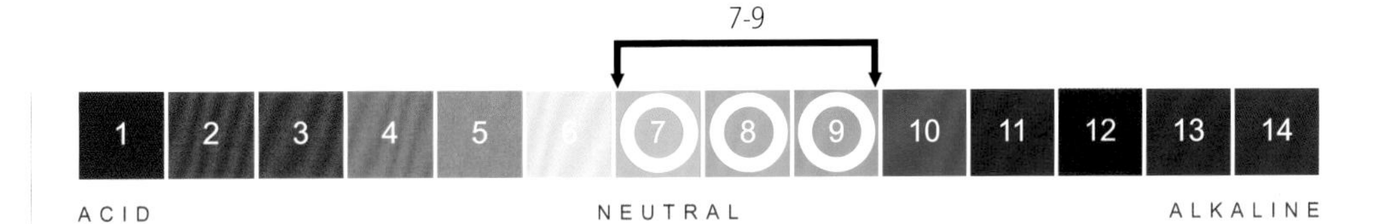

Temperature

52-74°C (126-165°F)

Description

Synechococcus species are rod-shaped cyanobacteria that create green mats and can form some of the most prominent green colors in thermal features. They are photosynthetic and intolerant of sulfur. They contain pigments (phycobiliproteins) that fluoresce orange red at wavelengths of 570-620 nm, which can serve to identify them. Part of a large community of microbes that live in mats, they are the main source of organic carbon that is available to neighboring organisms such as *Chloroflexus*. Like *Calothrix*, *Synechococcus* has two very different metabolisms – using photosynthesis during the day for energy and fermentation at night.

Habitat

Synechococcus is found around Yellowstone in neutral to alkaline springs that are non-sulfidic (do not smell like rotten eggs) such as Mammoth Hot Springs, Grand Prismatic Spring, Imperial Geyser and Octopus Spring. Species of *Synechococcus* are very prevalent in the oceans around the world and may play an important role in the global carbon cycle by transforming abundant amounts of carbon dioxide into oxygen.

Thermoproteus

UPPER, MIDWAY,
LOWER GEYSER BASINS

Thermoproteus

domain: **ARCHAEA**
phylum: **CRENARCHAEOTA**

pH

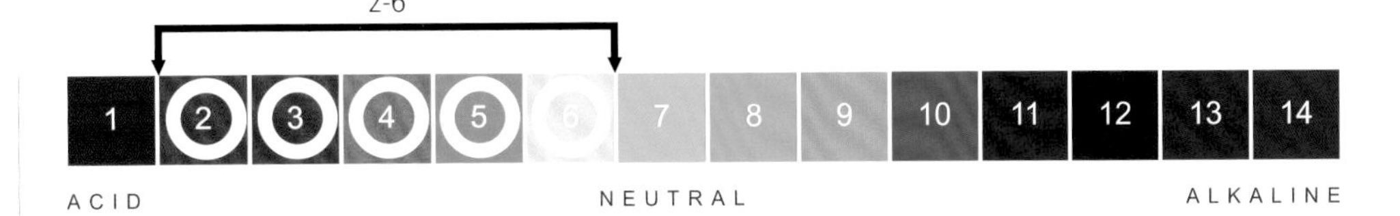

Temperature

70-100°C (158-212°F)

Description

Thermoproteus is part of the Archaea domain, and similar organisms are thought to be among some of Earth's earliest forms of life. It is shaped like a stiff rod. Species of *Thermoproteus* assimilate carbon dioxide into their bodies using inorganic sulfur and hydrogen as energy sources and produce hydrogen sulfide, which is harmful to human health.

Habitat

Species of *Thermoproteus* can be found in hot acidic springs in Yellowstone areas such as the Gibbon Geyser Basin, the Firehole River Drainage, and Mud Volcano. Species have also been found in hot springs in Iceland, Italy, North America, New Zealand, the Azores, and Indonesia.

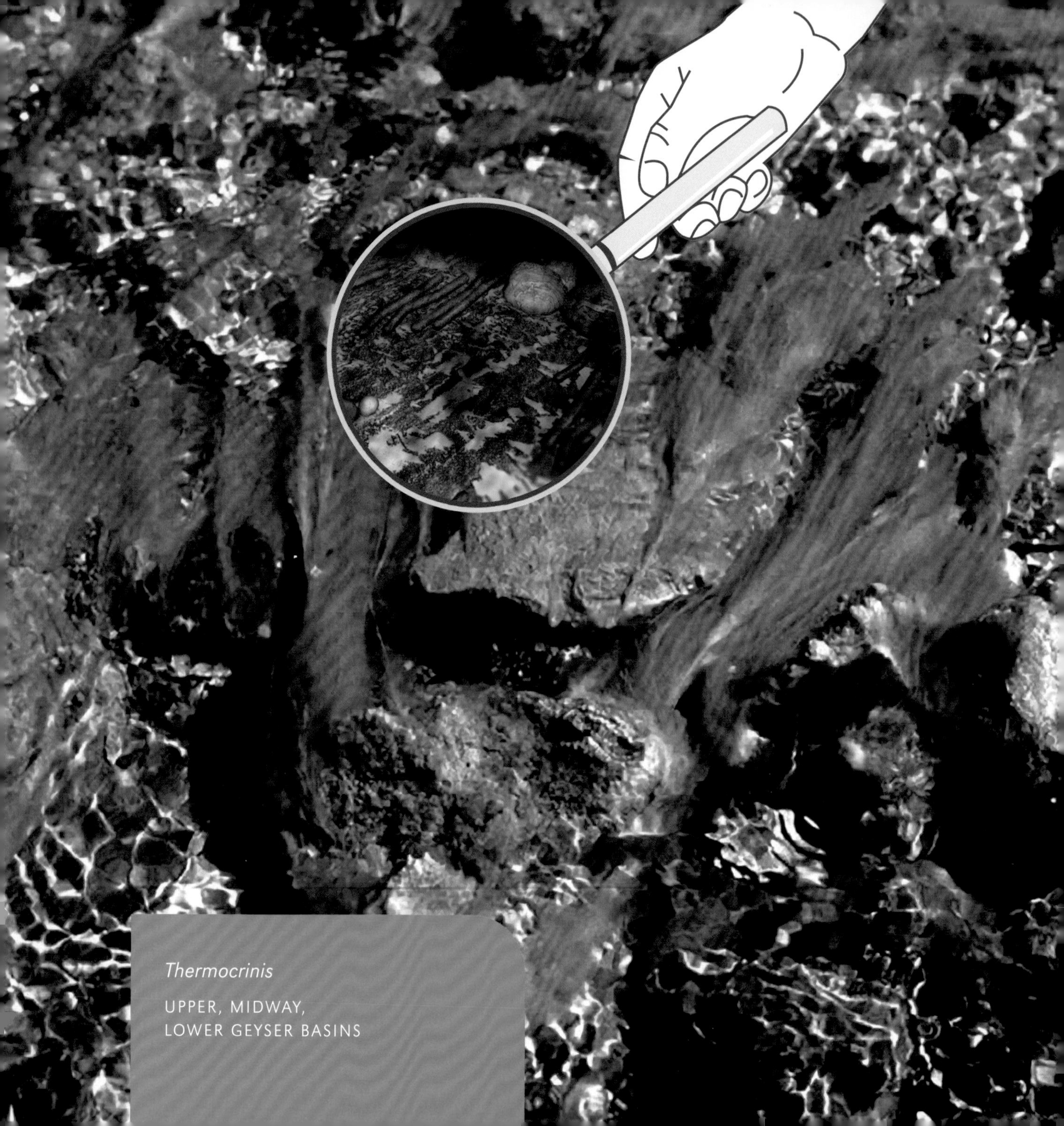

Thermocrinis

UPPER, MIDWAY,
LOWER GEYSER BASINS

Thermocrinis

domain: **BACTERIA**
phylum: **AQUIFICAE**

pH

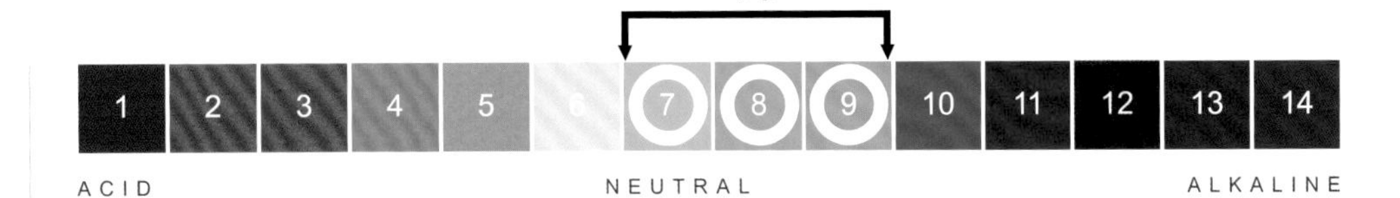

Temperature **55-91°C (131-195°F)**

Description

Thermocrinis is a rod-shaped bacterium that grows in the outflow of several alkaline hot springs in Yellowstone. Long chains of its cells form yellowish or pink streamers that attach to the sinter (a chemical crust) created by the precipitation of silicates in the water. It utilizes hydrogen gas (H_2) or sulfur with oxygen (O_2) to produce energy for metabolism and uses carbon dioxide (CO_2) as a carbon source.

Habitat

Thermocrinis is found in Yellowstone in many alkaline hot springs such as Octopus Spring in the Lower Geyser Basin. A similar species was found in a volcano in Costa Rica.

Thermus
UPPER, MIDWAY,
LOWER GEYSER BASINS

Thermus

domain: **BACTERIA**
phylum: **DEINOCOCCUS-THERMUS**

pH

Temperature

40-79°C (104-174°F)

Description

Thermus is a rod-shaped bacterium that sometimes forms bright red or orange streamers. It contains pigments called cartenoids that act as a sunscreen and protect it from high levels of sunlight. For energy, *Thermus* uses organic compounds from surrounding soils or other living or dead microbes in its environment. Discovered in 1968, it was one of the first extremophiles found in Yellowstone. The species *Thermus aquaticus* was the original source material for the enzyme *Taq* polymerase that is harvested and used in the Polymerase Chain Reaction (PCR), which allows scientists to make many copies of DNA. This process is used to identify organisms or individuals for biodiversity studies and in crime scene investigations. PCR is so important that Kary Mullis won a Nobel Prize for developing it.

Habitat

Thermus can be found around Yellowstone in thermal areas such as those around Firehole Lake Drive and Octopus Spring. *Thermus* species have also been found in deep-sea hydrothermal vents; hot springs in California and Iceland; and even in hot-water heaters in residential homes.

GRAND PRISMATIC

1

Grand Prismatic Spring is the largest hot spring in America and one of the most photographed features of Yellowstone. Its temperature ranges from 63°C to 87°C (145°F to 188°F). It has about the same alkalinity as baking soda with a pH of 8.3.

The vivid blue water in the center is the hottest part of the spring. Many people believe life cannot exist in the center of the spring, but that area often contains a small population of microbes that use chemicals as a source of energy.

Calothrix is a cyanobacterium that is dark brown in color. It lives in water above 30°C (86°F). *Calothrix* contains a chemical that acts like a sunscreen and protects it from high levels of UV radiation. **(1)**

Synechococcus, a type of bacterium, is a greenish oxygenic photosynthesizer that can live in temperatures as high as 74°C(165°F). **(2)**

Phormidium is an orange cyanobacterium that can form long streamers or layered sedimentary structures called stromatolites. It exists in a temperature range from 35-57°C (95°F-135°F). **(3)**

Excelsior Geyser was once the largest geyser in the world, but its last major eruptions were during the 1880s. Its water is 55°C (199°F) with a pH of 8.4. **(4)**

As the water temperature drops to approximately 70°C (158°F), photosynthetic bacteria form microbial mats ranging from green to red. In the winter, mats are often green. In the summer, the mats often turn orange and red, because they produce carotenoids, which act as a sunscreen, to protect them from harmful sunlight. **(5)**

MAMMOTH

Sulfur-oxidizing bacteria like *Thermochromatium* form deep purple-red or reddish-brown mats. Temperature range is 34-57°C (93-135°F). **(1)**

Bacteria such as *Chloroflexus* and *Chlorobium* entwine into mats with other thermophiles. They perform photosynthesis, and similar species may have been among the first on the Earth to do so. **(2)**

Mt. Everts (2387m, 7831ft) consists of sedimentary rocks deposited by a shallow inland sea 70 to 140 million years ago. **(3)**

pH 6.3 Temperature 54°C (129°F) **(4)**

Cyanobacteria such as *Oscillatoria* are found in many of the **Mammoth Hot Springs** features. They can range from green to orange and brown but are usually orange. They usually form thick mats embedded with calcium deposits. *Oscillatoria* can move in response to light and temperature changes. Temperature range is 36-45°C (96-104°F). **(5)**

Green nonsulfur bacteria such as *Chloroflexus* and *Chlorobium* mix with cyanobacteria to create mats that change color depending on the amount of sunlight. They are usually dark green in the winter and light, green yellow in the summer because of pigments called carotenoids that protect the organisms from damaging light much like sunscreen. **(1)**

Oscillatoria are orange cyanobacteria that are named for the slow back and forth oscillations they make in response to light and temperature changes. **(2)**

Diatoms, microscopic single-celled algae, mix with bacteria to form a mat. These mats also can host protozoa like *Chilodonella* that eat bacteria and diatoms. **(3)**

Orange Spring Mound pH 6.7 Temperature 45°C (113°F)

When water stops flowing over the terraces, the microbial communities die, and only the gray travertine is visible. **Mammoth** is very active, and the water flow of the springs is constantly changing. Microbes that thrive at different temperatures have different pigments, so as the spring water changes temperature, the colors of the microbes change too.

Cyanidioschyzon pH 0-4

Temperature range 40-55°C (104-131°F). *Cyandioschyzon* is one of the most heat and acid-tolerant algae known. It forms a bright green coating or mat on top of orange-red iron deposits in runoff channels. **(1)**

Pinwheel Geyser pH 3.42

Temperature 42.8°C (109°F) **(2)**

Whirligig Geyser pH 3.4

Whirligig goes through active and dormant periods. Its eruptions last 3 to 5 minutes and can reach from 1.5 to 4.5 meters (5 to 15 feet) in height. Whirligig's water is much hotter than other nearby geysers. The heat prevents *Cyanidioschyzon* from growing, but another type of thermophile, *Metallosphaera,* thrives by oxidizing the abundant iron in the water, forming the orange community. **(3)**

Iron oxides and arsenic create the red-orange colors in Whirligig Geyser's runoff. The changes in colors from oranges to greens are due to temperature and chemical gradients in the water channels, allowing various microbial communities to form mats. **(4)**

Constant Geyser pH 3.46

Temperature 37.8°C (100°F)

This geyser has frequent eruptions that can go 6 to 9 meters (20-30 feet) high but last barely 10 seconds. **(5)**

Zygogonium pH 0-4

Temperature Range 32-55°C (90-131°F). This mildly thermophilic alga is photosynthetic and turns dark purple or black in bright sunlight. Its cylinder-shaped cells join end-to-end to make long filaments, mats and streamers. **(6)**

Little Whirligig pH 3.29

Temperature 26.7°C (80°F). This geyser is much cooler than its neighbors and has been dormant for several years. Orange-yellow iron oxide deposits are visible. **(7)**

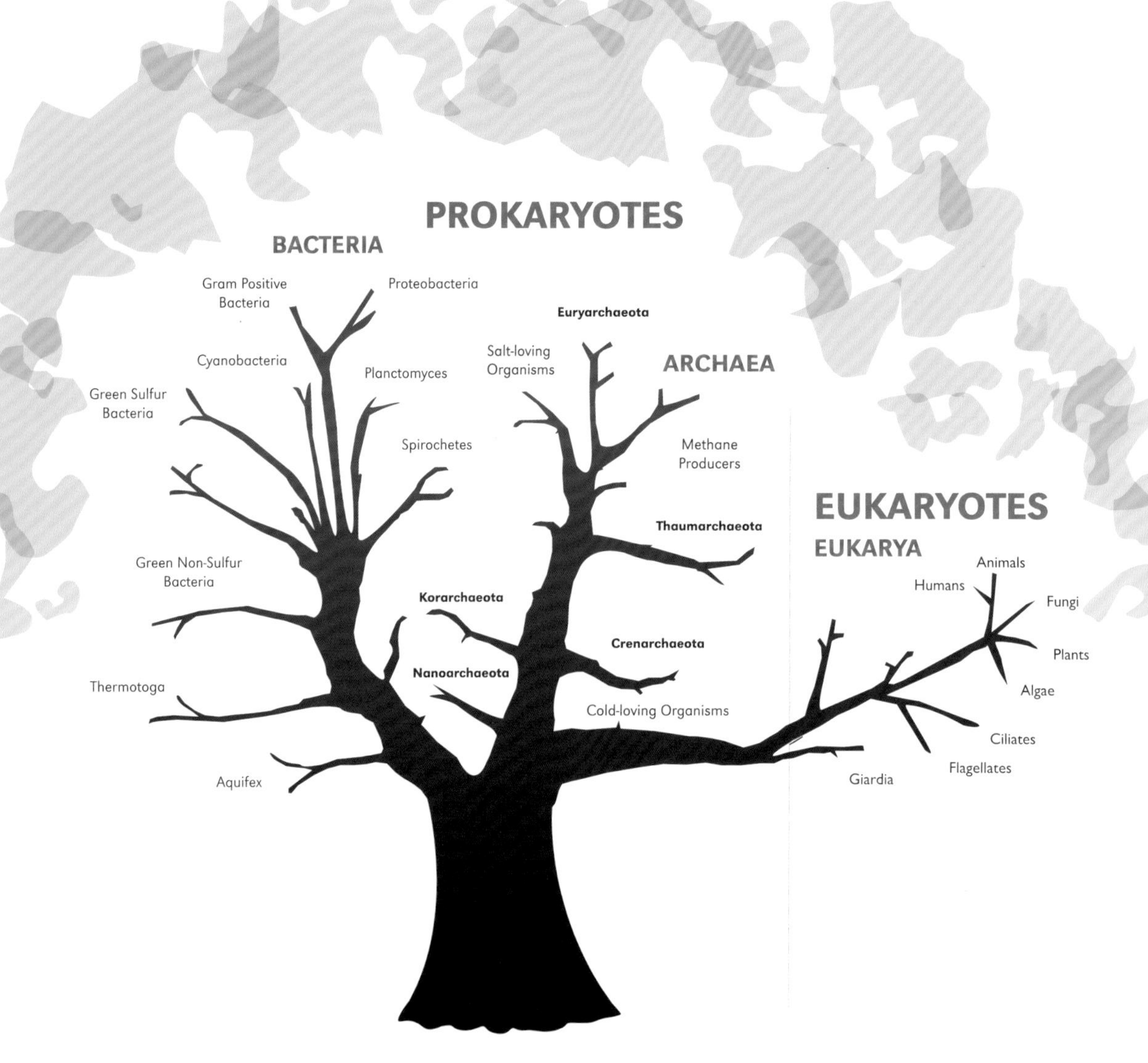
PROKARYOTES
BACTERIA
Gram Positive Bacteria
Proteobacteria
Cyanobacteria
Planctomyces
Green Sulfur Bacteria
Spirochetes
Green Non-Sulfur Bacteria
Thermotoga
Aquifex
Euryarchaeota
Salt-loving Organisms
ARCHAEA
Methane Producers
Thaumarchaeota
Korarchaeota
Crenarchaeota
Nanoarchaeota
Cold-loving Organisms
EUKARYOTES
EUKARYA
Animals
Humans
Fungi
Plants
Algae
Ciliates
Flagellates
Giardia

Tree of Life

Life, as we understand it today, is divided into three major domains: Eukarya, Bacteria and Archaea. The divisions reflect differences in fundamental ways in which these organisms live and their unique evolutionary histories.

Many biology students are familiar with the distinction between eukaryotes and prokaryotes. Eukaryotes are thought to have arisen later in the evolution of life. They have complex cellular structure – a nucleus and membrane-bound organelles, and thus a division of labor within the cell. All multicellular life is eukaryotic. All animals and plants are eukaryotes. But many microbes are also eukaryotes, including yeast, paramecia, microalgae, and the protists (unicellular organisms that do not fit into other kingdoms). In general, eukaryotes are larger, more complex and less adapted to extreme environments that prokaryotes.

The term prokaryotes is used to define single-celled organisms that lack a membrane-bound nucleus. However, prokaryote is an overly broad term that encompasses organisms that are as different from each other as ***E.coli*** is from humans. Prokarya is divided into two major domains, Bacteria and Archaea. Distinctions, it seemed, were primarily based on habitat, with Archaea generally occupying the more extreme environments and Bacteria the more moderate climes. However, with advances in technology, scientists found the presence of Archaea to be ubiquitous. They live in soils, in our guts, in the oceans, and, in fact, in almost every environment in which we look. Some Bacteria and Archaea thrive at temperatures up to 113°C (252°F). However, Archaea tend to be more dominant in extreme environments while Bacteria are the major players in the more moderate environments. The name Archaea derives from the word "archae," which means ancient, because Archaea are thought to be similar to some of first forms of life on Earth and many live in conditions similar to those found on early Earth (extremely hot with little or no oxygen).

pH

Concentration of hydrogen ions compared to distilled water		Examples of solutions at this pH
10,000,000	ph=1	Battery acid
1,000,000	ph=2	Stomach acid (Gastric juices)
100,000	ph=3	Lemon juice, Vinegar
10,000	ph=4	Grapefruit, Orange juice, Soda, Apples
1,000	ph=5	Acid rain, Tomato juice, Wine, Beer
100	ph=6	Rain water, Black coffee
10	ph=7	Urine, Saliva, Milk
1	ph=8	"Pure" water, Human blood
1/10	ph=9	Sea water, Egg whites
1/100	ph=10	Baking soda
1/1,000	ph=11	Milk of magnesia, Tums ™ antacid
1/10,000	ph=12	Ammonia, Detergents, Fertilizer
1/100,000	ph=13	Bleach, Lime
1/1,000,000	ph=14	Oven cleaner
1/10,000,000		Liquid drain cleaner

In Yellowstone National Park, the pH of the thermal features varies dramatically and is one of the key factors that determine which microbial communities are present.

pH is a measure of the acidity or basicity of water. Pure water is said to be neutral, with a pH of 7. Solutions with a pH less than 7 are acidic, and solutions with a pH greater than 7 are basic or alkaline. The pH scale is logarithmic, which means that as the pH moves away from neutral, each increment is a 10-fold change, such that pH 6 is 10 times more acidic than pH 7. pH 5 is 10 times more acidic than pH 6, and 100 times more acidic than pH 7, and so on. Norris Geyser Basin is one of the hottest and most acidic locations in Yellowstone and will give you a unique look at extreme environments. While in the Upper Geyser Basin, which includes Old Faithful, there are several alkaline springs that have a pH greater than 9. pH across the Yellowstone ecosystem varies from below 1 to nearly 10, representing a 1-billion-fold difference in acidity across the Park.

Temperature Limits

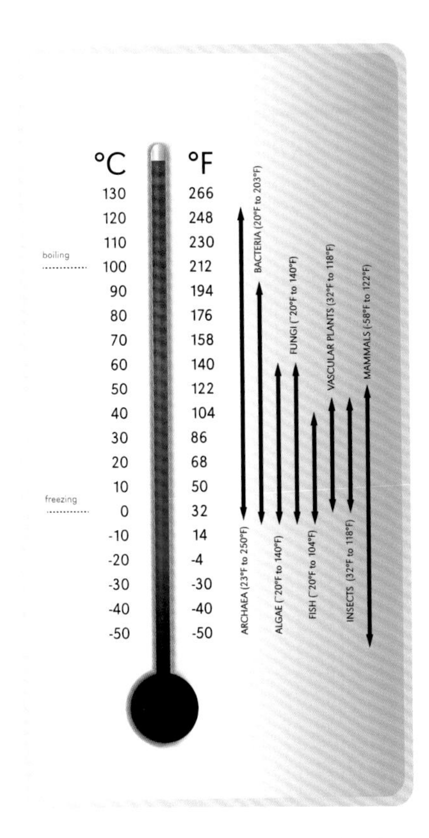

This chart reflects the temperature extremes that these organisms may be able to withstand, and not their optimal temperature range.

Microbiology has long focused on the organisms with medical relevance. Microbes that can infect us and other vertebrates thrive in the temperature in which the host organism lives: for humans this temperature is around 37°C (98.6°F).

Few scientists thought that life could exist in the extremes of temperature and other parameters that exist on the planet – for example, the extreme heat in Yellowstone or the cold in Antarctica. However, a closer look at these environments reveals that they not only support life, many organisms actually thrive there. We now know that life can survive at temperatures up to and exceeding 113°C (235°F). This is remarkable, considering that 121°C (250F), considered the 'gold standard' of sterilization, has been used to kill all previously known microbes. Most organisms are literally cooked at these temperatures, with their proteins denaturing, much like a cooked egg; lipids and other molecules become unstable and melt. Understanding how these organisms thrive at such extremes can help us understand how life began on Earth. Yellowstone National Park's thermal features are a great example of the variety of complex communities that thrive in extreme environments.

Living Colors

A spectrum of colors is on display in Yellowstone's many thermal features. The variety of colors you see may be due to the presence of minerals or microbes, but often is determined by the presence of both. How can you determine whether something you are looking at is predominately minerals or microbes without using a microscope? You will not be able to tell definitively, but you can make an educated guess by taking the following into account, but remember both minerals and microbes are most likely present.

- Minerals tend to deposit around the edge of the pools, yet microbes grow out from the source of the pools and along runoff channels.
- Mineral deposits look crystalline or hard, whereas microbes look spongy, soft or wavy.

Microbial colors can act as a living thermometer. Millions of tiny microbes will bond together into groups and form filaments or mats that are often distinct colors. The colors indicate where the water changes temperature, therefore providing a giant map of temperature gradients or occasionally chemical or pH gradients.

Microbial mats can be delicate and thin like a piece of tissue paper or thick and slimy like a piece of soggy lasagna. Mats can consist of many layers of organisms mixed together and interacting to form a complex interdependent community. Over time mats can grow to have multiple layers and be over an inch thick.

Microbial mats are tiny ecosystems that have layers much like a forest. Organisms living at the top of the mat use sunlight for energy and perform photosynthesis, much like a forest canopy. Organisms lower in the mat use chemicals produced by the microbes in the upper layer to get energy. They also recycle nutrients and help with functions such as decomposition, just like some of the organisms in a forest's understory.

Microbial communities are constantly changing. Variables such as the amount of sunlight or water can dramatically change the appearance of a thermal feature. Hot springs can transform into mud pots or fumaroles, or dry up altogether. Geysers can stop erupting, or new hot springs can emerge. Seasonal changes in sunlight can alter the colors of a thermal feature as microbes sensitive to light recede from the surface, or microbes that get energy from sunlight utilize dark pigments to protect them like sunscreen.